Building and Sending Interstellar Probes

Building and Sending Interstellar Probes

Building and Sending Interstellar Probes

Other books by Martin K. Ettington

Spiritual and Metaphysics Books:
Prophecy: A History and How to Guide
God Like Powers and Abilities
Enlightenment for Newbies
Removing Illusions to Find True
    Happiness
Using the Scientific Method to Study
    the Paranormal
A Compendium of Metaphysics and
    How to Guides (Six books
    together in one volume)
Love from the Heart
The Enlightenment Experience
Learn Your Soul's Purpose
Pursuing Enlightenment
A Modern Man's Search for Truth
Use Intuition and Prophecy to Improve
    Your Life
The Handbook of Spiritual and Energy
    Healing

Longevity & Immortality:
Physical Immortality: A History and
    How to Guide
The Commentaries of Living Immortals
Records of Extremely Long Lived
    Persons
Enlightenment and Immortality
Longevity Improvements from Science
The 10 Principles of Personal
    Longevity
Telomeres & Longevity
The Diets and Lifestyles of the Worlds
    Oldest Peoples
The Longevity Six Books Bundle

Science Fiction:
Out of This Universe
Personal Freedom-Parts 1 & 2
The Psychic Soldier Series:
    Book 1-Himalayan Journey
    Book 2-A Soldier is Born
    Book 3-Fighting For Right
    Book 4-Earth Protector
The Immortality Sci Fi Bundle

The God Like Powers Series:
Human Invisibility
Invulnerability and Shielding
Teleportation
Psychokinesis
Our Energy Body, Auras, and
    Thoughtforms

The God Like Powers Series—
    Volume 1 Compilation

The Yoga Discovery Series:
Yoga-An Ancient Art Form
Hatha Yoga-Helping you Live Better
Raja Yoga-Through the Ages
The Yoga Discovery Package

Business & Coaching Books:
Creating, Publishing, & Marketing
    Practitioner Ebooks
Building a Successful Longevity
    Coaching Business
Why Become a Coach?
The Professional Coaching Success
    Trilogy
2020-Make Money Writing and Selling
    Books
The 2020 Handbook of High Paying
    Work Without a College Degree

Science, Technology, and Misc.
Future Predictions By and Engineer &
    Seer
The Unusual Science & Technology
    Bundle
The Real Atlantis-In the Eye of the
    Sahara
Are Cryptozoological Animals Real or
    Imaginary?
Real Time Travel Stories From a
    Psychic Engineer
Removing Limits On Our
    Consciousness-And
    Thinking Outside the Box
33 Incredible True Survival Stories
How to Survive Anything: From the
    Wilderness to Man Made
    Disasters
All About Mars Journeys and
    Settlement
Mining the Asteroid Belt

Ancient History
The Real Atlantis-In the Eye of the
Sahara
Ancient & Prehistoric Civilizations
Ancient & Prehistoric Civilizations-Book
    Two
The History of Antediluvian Giants
The Antediluvian History of Earth
Ancient Underground Cities and
    Tunnels
Strange Objects Which Should Not Exist
Strange and Ancient Places in the USA
A Theory of Ancient Prehistory And
    Giant Aliens
Aliens and Space

Building and Sending Interstellar Probes

Aliens and Secret Technology
Aliens Are Already Among Us
Designing and Building Space Colonies
Humanity and the Universe
All About Moon Bases
All About Mars Journeys and Settlement

The Space and Aliens Six Books Bundle
A Theory of Ancient Prehistory and
 Giant Aliens
The Space Colonies and Space
 Structures Coloring Book
All About Asteroids

The Longevity Training Series

(A transcription of the online Multimedia Longevity Coaching Training Program)

The Personal Longevity Training Series-Book1-Long Lived Persons
The Personal Longevity Training Series-Book2-Your Soul's Purpose
The Personal Longevity Training Series-Book3-Enable Your Life Urge
The Personal Longevity Training Series-Book4-Your Spiritual Connection
The Personal Longevity Training Series-Book5-Having Love in Your Heart
The Personal Longevity Training Series-Book6-Energy Body Health
The Personal Longevity Training Series-Book7-The Science of Longevity
The Personal Longevity Training Series-Book8-Physical Body Health
The Personal Longevity Training Series-Book9-Avoiding Accidents
The Personal Longevity Training Series-Book10-Implementing These Principles

The Personal Longevity Training Series-Books One Thru Ten

These books are all available in digital and printed formats from my
website and on Amazon, Barnes & Noble, Apple ITunes, and many other sites

My Books Website is: http://mkettingtonbooks.com

Building and Sending Interstellar Probes

Man has dreamed of travelling to the stars for as long as he knew that those points of light in the sky were stars like the Sun.

Now, in the 21st century there are actual studies and plans to send interstellar space probes to the nearest stars. Probably to Alpha Centauri which is only 4.3 light years away.

What is the history of plans to get to the nearest star? And what are the plans today to make these ideas a reality?

In this book we explore the history of the scientific effort to send "Star Probes" and how they will be built.

Plans for building interstellar probes are discussed and the technology we need to build to get there is reviewed too.

How will we be able to send them at a significant fraction of the speed of light to get there in many persons lifetimes? And what technologies will we need to develop to make this happen?

This is the start of an amazing true life adventure.

Building and Sending Interstellar Probes

Signup for our Mailing List to get the following:

1) A discount coupon for 25% discount on all books on our site

2) Occasional Notices of new books available

3) Occasional Email on other offerings of ours (Monthly)

Go to this link to sign-up:

http://personal-longevity.com/mkebooks/emailsignup/

And click this link to get the FREE 102 page Ebook titled "Secrets of Many Things"

If you have any questions about this book or other subjects please contact the Author at:

mke@mkettingtonbooks.com

Building and Sending Interstellar Probes

Building and Sending Interstellar Probes

Table of Contents

Building and Sending Interstellar Probes

Building and Sending Interstellar Probes

1.0 Introduction

This book is number 5 in "The Living in Space Series". The other books covered topics like space colony designs, moon bases, traveling and living on Mars, and mining the Asteroids.

This book's topic is a little different since it covers the plans and thinking about building and sending an interstellar probe to the nearest star—Alpha Centauri

Plans for building interstellar probes are discussed and the technology we need to build them to get there is reviewed too.

Some of our spacecraft like the Voyager probes have already left the Solar System, but new technologies will need to be developed to send probes at a significant fraction of the speed of light to get there within most people's lifetimes.

This calls for new thinking, new ideas, and some ambition to make it all happen.

Visiting the stars is of course one of man's oldest dreams, and we are living in the era which will start these exciting explorations.

Building and Sending Interstellar Probes

Building and Sending Interstellar Probes

2.0 What Does Interstellar Mean?

By definition an interstellar probe would travel to outside of the Sun's heliopause at least. It's likely that this mission would also be targeted to visit the nearest stars.

<u>The Closest Stars</u>

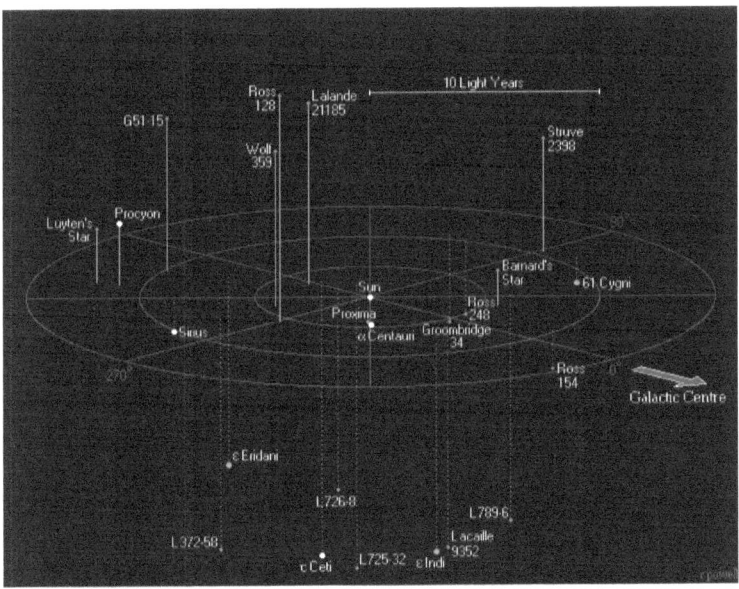

The above diagram shows the closest stars to the Sun within 10 light years in all directions. Of course the closest star is Proxima Centauri which is 4.244 light years away from us. It is also a trinary system so there are three stars in close proximity. The fact that it is our nearest star companion is the reason it would be a likely target for our first space probe to another star.

Proxima Centauri is a small, low-mass star located 4.244 light-years (1.301 pc) away from the Sun in the southern constellation of Centaurus. Its Latin name means the "nearest [star] of Centaurus". This object was discovered in 1915 by Robert Innes and is the nearest-known star to the Sun. With a quiescent apparent magnitude of 11.13, it is too faint to be seen with the naked eye.

Proxima Centauri is a member of the Alpha Centauri system, being identified as component **Alpha Centauri C**, and is 2.18° to the southwest of the Alpha Centauri AB pair. It is currently 12,950 AU (1.94 trillion km) from AB, which it orbits with a period of about 550,000 years.

A planet around the star

In 2016, astronomers announced the discovery of Proxima Centauri b, a planet orbiting the star at a distance of roughly 0.05 AU (7.5 million km) with an orbital period of approximately 11.2 Earth days.

According to updated measurements by the ESPRESSO spectrograph, its estimated mass is at least 1.17 times that of the Earth. The equilibrium temperature of Proxima b is estimated to be within the range of where water could exist as liquid on its surface, thus placing it within the habitable zone of Proxima Centauri, although because Proxima Centauri is a red dwarf and a flare star, whether it could support life is disputed.

Building and Sending Interstellar Probes

3.0 Why Send Interstellar Probes?

Man has always wanted to explore the stars to see if they have life or even intelligent life. Sure, we need to explore and settle the Solar System first, but it's in our genes that we need to see what the nearest star is like and explore any planets there.

Man wants to settle new places and there are no locations which are livable like Earth in our Solar System. Finding another survivable planet humanity could also live on would be a great accomplishment. There are many many science fiction stories about travelling to and settling planets around other stars. I've written a couple of those books myself.

We will probably also want to send any probes at as close to light speed as possible so it will reach there in the memories of people living when it was launched.

Building and Sending Interstellar Probes

Building and Sending Interstellar Probes

4.0 A History of Interstellar Probes

You might be surprised to learn that a number of probes launched over forty years ago are now in interstellar space. However, due to their slow speeds they would not reach another star for thousands of years. We obviously need something much faster.

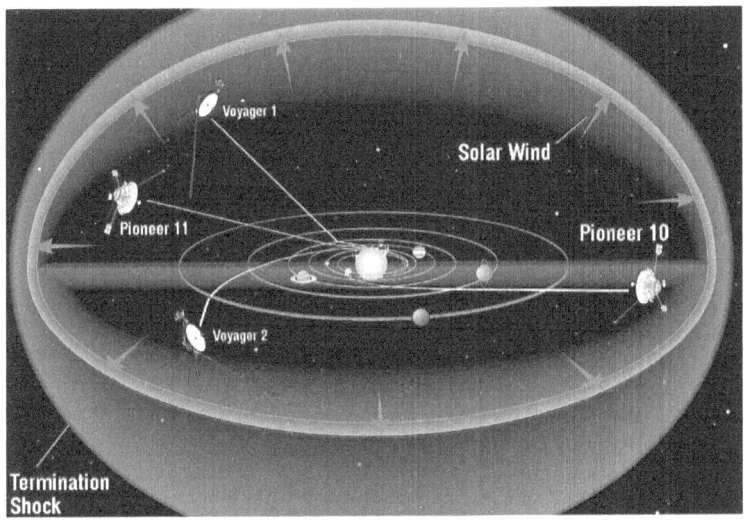

Probes which have already left the Solar System

4.1 Exploration Probes Already Sent

An interstellar probe is a space probe that has left—or is expected to leave—the Solar System and enter interstellar space, which is typically defined as the region beyond the heliopause. It also refers to probes capable of reaching other star systems.

There are five interstellar probes, all launched by the American space agency NASA: Voyager 1, Voyager 2, Pioneer 10, Pioneer 11 and New Horizons. As of 2019,

Building and Sending Interstellar Probes

Voyager 1 and Voyager 2 are the only probes to have actually reached interstellar space. The other three are on interstellar trajectories.

The termination shock is the point in the heliosphere where the solar wind slows down to subsonic speed. Even though the termination shock happens as close as 80–100 AU (Astronomical unit) the maximum extent of the region in which the Sun's gravitational field is dominant (the Hill sphere) is thought to be at around 230,000 astronomical units (3.6 light-years). This point is close to the nearest known star system, Alpha Centauri, located 4.36 light years away. Although the probes will be under the influence of the Sun for a long time, their velocities far exceed the Sun's escape velocity, so they will eventually leave forever.

Functioning Spacecraft

Voyager 1 (1977+)

Voyager 1 is a space probe launched by NASA on September 5, 1977. At a distance of about 148.42 AU (2.220×1010 km) as of 14 July 2020, it is the farthest manmade object from Earth.

It was later estimated that Voyager 1 crossed the termination shock on December 15, 2004 at a distance of 94 AU from the Sun.

At the end of 2011, Voyager 1 entered and discovered a stagnation region where charged particles streaming from the Sun slow and turn inward, and the Solar System's magnetic field is doubled in strength as interstellar space appears to be applying pressure. Energetic particles originating in the Solar System declined by nearly half, while the detection of high-energy electrons from outside increases 100-fold. The inner edge of the stagnation region is located approximately 113 astronomical units. (AU) from the Sun.

In 2013 it was thought Voyager 1 crossed the heliopause and entered interstellar space on August 25, 2012 at distance of 121 AU from the Sun, making it the first known human-manufactured object to do so.

As of 2017, the probe was moving with a relative velocity to the Sun of about 16.95 km/s (3.58 AU/year).

If it does not hit anything, Voyager 1 could reach the Oort cloud in about 300 years

Voyager 2 (1977+)

Plot of Voyager 2's heliocentric velocity against its distance from the sun, illustrating the use of gravity assist to accelerate the spacecraft by Jupiter, Saturn and Uranus. The spacecraft's encounter with Neptune actually decelerated the probe because of the way it encountered the planet.

Voyager 2 crossed the heliopause and entered interstellar space on November 5, 2018. It had previously passed the termination shock into the heliosheath on October 30, 2007. As of 14 July 2020 Voyager 2 is at a distance of 123.12 AU (1.842×1010 km) from Earth. The probe was moving at a velocity of 3.25 AU/year (15.428 km/s) relative to the Sun on its way to interstellar space in 2013.

It's moving at a velocity of 15.4 km/s (55,000 km/h) relative to the Sun as of December 2014. Voyager 2 is expected to provide the first direct measurements of the density and temperature of the interstellar plasma.

New Horizons (2006+)

New Horizons was launched directly into a hyperbolic escape trajectory, getting a gravitational assist from Jupiter en route. By March 7, 2008, New Horizons was 9.37 AU from the Sun and traveling outward at 3.9 AU per year. It will, however, slow to an escape velocity of only 2.5 AU per year as it moves away from the Sun, so it will never catch up to either Voyager. As of early 2011, it was traveling at 3.356 AU/year (15.91 km/s) relative to the Sun. On July 14, 2015, it completed a flyby of Pluto at a distance of about 33 AU from the Sun. New Horizons next encountered 486958 Arrokoth on January 1, 2019, at about 43.4 AU from the Sun.

The Heliosphere's termination shock was crossed by Voyager 1 at 94 astronomical units (AU) and Voyager 2 at 84 AU according to the IBEX mission.

If New Horizons can reach the distance of 100 AU, it will be traveling at about 13 km/s (29,000 mph), around 4 km/s (8,900 mph) slower than Voyager 1 at that distance.

Building and Sending Interstellar Probes

4.2 Past Proposed Interstellar Projects

There have actually been quite a few plans and projects developed to launch an interstellar probe to visit other stars. Here are some of those projects:

Project Orion (1958–1965)

Project Orion was a proposed nuclear pulse propulsion craft that would have used fission or fusion bombs to apply motive force. The design was studied during the 1950s and 1960s in the United States of America, with one variant of the craft capable of interstellar travel.

Bracewell probe (1960)

Interstellar communication via a probe, as opposed to sending an electromagnetic signal.

Sanger Photon Rocket (1950s-1964)

Eugene Sanger proposed a spacecraft powered by antimatter in the 1950s. Thrust was intended to come from reflected gamma-rays produced by electron-positron annihilation.

Enzmann Starship (1964/1973)

Proposed by 1964 and examined in an October 1973 issue of Analog, the Enzmann Starship proposed using a 12,000 ton ball of frozen deuterium to power thermonuclear powered pulse propulsion. About twice as long as the Empire State Building and assembled in-orbit, the spacecraft was part of a larger project preceded by large interstellar probes and telescopic observation of target star systems.

Building and Sending Interstellar Probes

Project Daedalus (1973–1978)

Project Daedalus was a proposed nuclear pulse propulsion craft that used inertial confinement fusion of small pellets within a magnetic field nozzle to provide motive force. The design was studied during the 1970s by the British Interplanetary Society, and was meant to flyby Barnard's Star in under a century from launch. Plans included mining Helium-3 from Jupiter and a pre-launch mass of over 50 thousand metric tons from orbit.

Starwisp (1985)

Starwisp is a hypothetical unmanned interstellar probe design proposed by Robert L. Forward. It is propelled by a microwave sail, similar to a solar sail in concept, but powered by microwaves from an artificial source.

Project Longshot (1987–1988)

Project Longshot was a proposed nuclear pulse propulsion craft that used inertial confinement fusion of small pellets within a magnetic field nozzle to provide motive force, in a manner similar to that of Project Daedalus. The design was studied during the 1990s by NASA and the US Naval Academy. The craft was designed to reach and study Alpha Centauri.

Medusa (1990s)

Medusa was a novel spacecraft design, proposed by Johndale C. Solem, using a large lightweight sail (spinnaker) driven by pressure pulses from a series of nuclear explosions. The design, published by the British Interplanetary Society, was studied during the 1990s as a means of interplanetary travel.

Building and Sending Interstellar Probes

Starseed launcher (1996)

Starseed launcher was concept for launching microgram interstellar probes at up to 1/3 light speed.

AIMStar (1990s-2000s)

AIMStar was a proposed antimatter catalyzed nuclear pulse propulsion craft that would use clouds of antiprotons to initiate fission and fusion within fuel pellets. A magnetic nozzle derived motive force from the resulting explosions. The design was studied during the 1990s by Penn State University. The craft was designed to reach a distance of 10,000 AU from the Sun in 50 years.

Project Icarus (2009+)

Project Icarus is a theoretical study for an interstellar probe and is being run under the guidance of the Tau Zero Foundation (TZF) and the British Interplanetary Society (BIS), and was motivated by Project Daedalus, a similar study that was conducted between 1973 and 1978 by the BIS. The project is planned to take five years and began on September 30, 2009.

Project Dragonfly (2014+)

The Initiative for Interstellar Studies (i4is) has initiated a project working on small interstellar spacecraft, propelled by a laser sail in 2014 under the name of Project Dragonfly. Four student teams worked on concepts for such a mission in 2014 and 2015 in the context of a design competition.

Geoffrey A. Landis proposed for interstellar travel future-technology project interstellar probe with supplying the energy from an external source (laser of base station) and ion thruster.

Interstellar Probe (ISP) (2018-)

A NASA funded study, led by the Applied Physics Laboratory, on possible options for an interstellar probe. The nominal concept would launch on a SLS in the 2030s. It would perform a powered Jupiter flyby or a very close perihelion and propulsive maneuver, and reach a distance of 1000-2000 AU within fifty years. Possibilities for planetary, astrophysical and exoplanet science along the way are also being investigated.

Interstellar Heliopause Probe (IHP) (2006)

A technology reference study published in 2006 with the ESA proposed an interstellar probe focused on leaving the heliosphere. The goal would be 200 AU in 25 years, with traditional launch but acceleration by a solar sail. The roughly 200–300 kg probe would carry a suite of several instruments including a Plasma Analyzer, Plasma radio wave experiment, Magnetometer, Neutral and charged atom detector, Dust analyzer, and a UV-photometer. Electrical power would come from an RTG.

NASA's Vision Mission; an early concept for the Innovative Interstellar Explorer

Solar frontier as envisioned at the turn of century, on a logarithmic scale (1999)

Building and Sending Interstellar Probes

Innovative Interstellar Explorer (2003)

NASA proposal to send a 35 kg science payload out to at least 200 AU. It would achieve a top speed of 7.8 AU per year using a combination of a heavy lift rocket, Jupiter gravitational assistance, and an ion engine powered by standard radioisotope thermal generators. The probe suggested a launch in 2014 (to take advantage of Jupiter gravitational assist), to reach 200 AU around 2044.

Realistic Interstellar Explorer and Interstellar Explorer (2000–2002)

Studies suggesting various technologies including Am-241-based RTG, optical communication (as opposed to radio), and low-power semi-autonomous electronics. Trajectory uses a Jupiter and Sun gravity assist to achieve 20 AU/year, allowing 1000 AU within 50 years, and a mission extension up to 20,000 AU and 1000 years. Needed technology included advanced propulsion and solar shield for perihelion burn around the Sun. Solar thermal (STP), nuclear fission thermal (NTP), and nuclear fission pulse, as well as various RTG isotopes were examined. The studies also included recommendations for a solar probe (see also Parker Solar Probe), nuclear thermal technology, solar sail probe, 20 AU/year probe, and a long-term vision of a 200 AU/year probe to the star Epsilon Eridani.

The "next step" interstellar probe in this study suggested a 5 megawatt fission reactor utilizing 16 metric tons of H2 propellant. Targeting a launch in the mid-21st century, it would accelerate to 200 AU/year over 4200 AU and reach the star Epsilon Eridani after 3400 years of travel in the year 5500 AD. However, this was a second-generation vision for a probe and the study acknowledged that even

20 AU/year might not be possible with then current (2002) technology. For comparison, the fastest probe at the time of the study was Voyager 1 at about 3.6 AU/year (17 km/s), relative to the Sun

TAU mission (1987)

TAU mission (Thousand Astronomical Units) was a proposed nuclear electric rocket craft that used a 1 MW fission reactor and an ion drive with a burn time of about 10 years to reach a speed of 106 km/s (about 20 AU/year) to achieve a distance of 1000 AU in 50 years. The primary goal of the mission was to improve parallax measurements of the distances to stars inside and outside our galaxy, with secondary goals being the study of the heliopause, measurements of conditions in the interstellar medium, and (via communications with Earth) tests of general relativity.

Building and Sending Interstellar Probes

5.0 Interstellar Technologies

Here are some technologies that have been discussed in relation to making an interstellar probe. These include the approaches to help us build a fast probe:

Gravity assist

A traditional gravity assist can be compared to throwing a tennis ball at a train (it rebounds not just with incoming velocity, but is accelerated by the train), it uses the gravity of a planet and its relative motion around the Sun compared to the spacecraft. For example, Voyager 2 increased its velocity by performing gravity assists at Jupiter, Saturn, and Uranus.

Oberth effect

In astronautics, a powered flyby, or Oberth maneuver, is a maneuver in which a spacecraft falls into a gravitational well and then accelerates as its falling, thereby achieving additional speed. The resulting maneuver is a more efficient way to gain kinetic energy than applying the same impulse outside of a gravitational well. The gain in efficiency is explained by the Oberth effect, wherein the use of an engine at higher speeds generates greater mechanical energy than use at lower speeds. In practical terms, this means that the most energy-efficient method for a spacecraft to burn its engine is at the lowest possible orbital periapsis, when its orbital velocity (and so, its kinetic energy) is greatest. In some cases, it is even worth spending fuel on slowing the spacecraft into a gravity well to take advantage of the efficiencies of the Oberth effect. The maneuver and effect are named after the person who first described them in 1927, Hermann Oberth, an Austro-

Hungarian-born German physicist and a founder of modern rocketry.

The Oberth effect is strongest at a point in orbit known as the periapsis, where the gravitational potential is lowest, and the speed is highest. This is because firing a rocket engine at high speed causes a greater change in kinetic energy than when fired at lower speed. Because the vehicle remains near periapsis only for a short time, for the Oberth maneuver to be most effective the vehicle must be able to generate as much impulse as possible in the shortest possible time. As a result the Oberth maneuver is much more useful for high-thrust rocket engines like liquid-propellant rockets, and less useful for low-thrust reaction engines such as ion drives, which take a long time to gain speed. The Oberth effect also can be used to understand the behavior of multi-stage rockets: the upper stage can generate much more usable kinetic energy than the total chemical energy of the propellants it carries.

The Oberth effect occurs because the propellant has more usable energy due to its kinetic energy in addition to its chemical potential energy. The vehicle is able to employ this kinetic energy to generate more mechanical power.

RTGs

An example of RTG used on a probe leaving the Solar system is the Voyagers. Typically these have used Plutonium but an RTG using 241Am was proposed for an interstellar type mission in 2002. This could support mission extensions up to 1000 years on the interstellar probe, because the power output would be more stable in the long-term than plutonium. Other isotopes for RTG were also examined in the study, looking at traits such as watt/gram, half-life, and decay products. An interstellar

probe proposal from 1999 suggested using three
advanced radioisotope power sources. An RTG using
241Am was also studied as RTG fuel by the ESA

<u>Ion engines</u>

An ion thruster or ion drive is a form of electric propulsion
used for spacecraft propulsion. It creates thrust by
accelerating ions using electricity.

An ion thruster ionizes a neutral gas by extracting some
electrons out of atoms, creating a cloud of positive ions.
These ion thrusters rely mainly on electrostatics as ions
are accelerated by the Coulomb force along an electric
field. Temporarily stored electrons are finally reinjected by
a neutralizer in the cloud of ions after it has passed
through the electrostatic grid, so the gas becomes neutral

again and can freely disperse in space without any further electrical interaction with the thruster. Electromagnetic thrusters on the contrary use the Lorentz force to accelerate all species (free electrons as well as positive and negative ions) in the same direction whatever their electric charge, and are specifically referred to as plasma propulsion engines, where the electric field is not in the direction of the acceleration.

Ion thrusters in operational use have an input power need of 1–7 kW (1.3–9.4 hp), exhaust velocity 20–50 km/s (45,000–112,000 mph), thrust 25–250 millinewtons (0.090–0.899 ozf) and efficiency 65–80% though experimental versions have achieved 100 kilowatts (130 hp), 5 newtons (1.1 lbf).

The Deep Space 1 spacecraft, powered by an ion thruster, changed velocity by 4.3 km/s (9,600 mph) while consuming less than 74 kg (163 lb) of xenon. The Dawn spacecraft broke the record, with a velocity change of 11.5 km/s (26,000 mph).

Applications include control of the orientation and position of orbiting satellites (some satellites have dozens of low-power ion thrusters) and use as a main propulsion engine for low-mass robotic space vehicles (such as Deep Space 1 and Dawn).

Ion thrust engines are practical only in the vacuum of space and cannot take vehicles through the atmosphere because ion engines do not work in the presence of ions outside the engine. Additionally, the engine's minuscule thrust cannot overcome any significant air resistance. Spacecraft rely on conventional chemical rockets to reach their initial orbit.

Building and Sending Interstellar Probes

Solar sails

Solar sails (also called light sails or photon sails) are a method of spacecraft propulsion using radiation pressure exerted by sunlight on large mirrors. Based on the physics, a number of spaceflight missions to test solar propulsion and navigation have been proposed since the 1980s.

A useful analogy to solar sailing may be a sailing boat; the light exerting a force on the mirrors is akin to a sail being blown by the wind. High-energy laser beams could be used as an alternative light source to exert much greater force than would be possible using sunlight, a concept known as beam sailing. Solar sail craft offer the possibility of low-cost operations combined with long operating lifetimes. Since they have few moving parts and use no propellant, they can potentially be used numerous times for delivery of payloads.

Solar sails use a phenomenon that has a proven, measured effect on astrodynamics. Solar pressure affects all spacecraft, whether in interplanetary space or in orbit around a planet or small body. A typical spacecraft going to Mars, for example, will be displaced thousands of kilometers by solar pressure, so the effects must be accounted for in trajectory planning, which has been done since the time of the earliest interplanetary spacecraft of the 1960s. Solar pressure also affects the orientation of a spacecraft, a factor that must be included in spacecraft design.

The total force exerted on an 800 by 800 meter solar sail, for example, is about 5 newtons (1.1 lbf) at Earth's distance from the Sun, making it a low-thrust propulsion system, similar to spacecraft propelled by electric engines, but as it uses no propellant, that force is exerted almost constantly and the collective effect over time is great enough to be considered a potential manner of propelling.

Building and Sending Interstellar Probes

6.0 Current Planning for Interstellar Missions

Scientists are quite active and already making initial plans for a probe mission to the nearest star….

Proposed Interstellar Mission Reaches for the Stars, One Generation at a Time

Starting in the early 2030s, the project could become our first purposeful step out of the solar system—if it launches at all

It would begin in the early 2030s, with a launch of a roughly half-ton nuclear-powered spacecraft on the world's largest rocket, designed to go farther and faster than any human-made object has ever gone before. The probe would pass by Jupiter and perhaps later dive perilously close to the sun, in both cases to siphon a fraction of each object's momentum, picking up speed to supercharge its escape. Then, with the sun and the major planets rapidly receding behind it, the craft would emerge from the haze of primordial dust that surrounds our star system, allowing it an unfiltered glimpse of the feeble all-sky glow from countless far-off galaxies. Forging ahead, it could fly by one or more of the icy, unexplored worlds now known to exist past Pluto. And gazing back, it could seek out the pale blue dot of Earth, looking for hints of our planet's life that could be seen from nearby stars.

All this would be but a prelude, however, to what McNutt and other mission planners pitch as the probe's core scientific purpose. About a decade after launch, it would pierce the heliosphere—a cocoonlike region around our solar system created by "winds" of particles flowing from our sun—to reach and study the cosmic rays and clouds of plasma that make up the "interstellar medium" that fills the dark spaces between the stars. Continuing its cruise, by the 2080s it could conceivably have traveled as far as 1,000 astronomical units (AU), or Earth-sun distances, from the solar system, achieving its primary objective at last: an unprecedented bird's-eye view of the heliosphere that could revolutionize our understanding of our place in the cosmos.

"We have seen heliospheres—'astrospheres'—around other stars, but we don't know the structure of our own," says Elena Provornikova, a researcher at APL leading planning for the mission's heliophysics science. "So imagine you are sitting inside your home, and someone asks you what it looks like from the outside. You'd need to step out to see. [Interstellar Probe] would be the first time we send a dedicated instrument payload to go outside and give us that picture."

Building and Sending Interstellar Probes

<u>Visiting An Astrosphere</u>

It would not be the first active spacecraft to go interstellar: NASA's Voyager 1 probe exited the heliosphere in 2012, followed by its twin, Voyager 2, in 2018. But the 1970s-era Voyager craft were accidental witnesses, designed for 4.5-year missions to study the planets of the outer solar system, not what lies beyond. Only a great deal of luck and ingenuity allowed them to survive to reach the heliosphere at all—with most of their vintage onboard instruments disabled for lack of power.

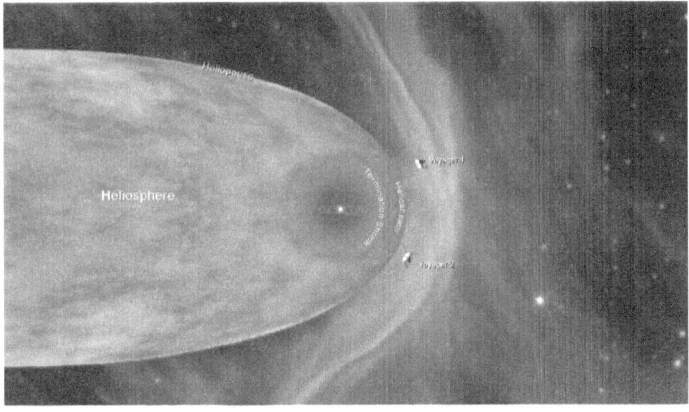

An artist's concept showing the locations of the Voyager probes as well as the notional architecture of the heliosphere. Studies by the Interstellar Probe and other missions could reveal the heliosphere's true shape, settling decades of scientific speculation.

The Voyager missions' meager measurements have revealed a complex, dynamic structure, produced by the interplay of our sun's exhalations with the interstellar medium's stronger headwinds. Both first passed through a transitional "heliopause" boundary region where the pressures of the solar wind and the interstellar medium are in a delicate balance. Outside the heliopause, the territory remains mostly uncharted. Sculpted by galactic magnetic fields and showers of particles from the ancient, explosive deaths of massive stars, our heliosphere may most resemble a wind sock, stretching out a long, undulating "tail" in its wake as our solar system moves through the Milky Way. Conversely, it may be more like a simple bubble or something in between these two cases—a croissantlike shape with twin, tapered trailing lobes. No one yet knows. But what is certain is that the heliosphere shields us from most of the high-energy cosmic rays that fill our corner of the galaxy—rays that, striking our fragile planet, can alter the climate and even our own DNA, ripping apart the cellular machinery at the base of Earth's biosphere.

"All these connections somehow unite to form our habitable astrosphere," Provornikova says. In some sense, part of the answer to the mystery of life's emergence and endurance on one small planet around one unremarkable yellow star may be found almost inconceivably far away, out at the unknown edges of the solar system.

Voyager 1, the current record holder for fastest outbound spacecraft, is traveling at 3.6 AU per year. McNutt is confident Interstellar Probe could go at least twice as fast—a lower limit that would allow the spacecraft to only get a little more than a third of the way to the aspirational goal of 1,000 AU within 50 years of its launch. "Everyone would like a number that is bigger than that," McNutt says, adding, though, that a more modest speed would allow more time for scientific observations at any given target along the way.

"To put this in perspective, we could get to where Voyager 1 is now about a dozen years after launch instead of the

Building and Sending Interstellar Probes

nearly half-century Voyager 1 has taken to get there," says Michael Paul, a space systems engineer at APL and one of Interstellar Probe's co-investigators. Provided, that is, that Interstellar Probe ever launches at all.

Big Rockets, Bigger Risks

First proposed in a 1958 report from the National Academy of Sciences, something like Interstellar Probe has consistently eluded space scientists, who have been stymied by shortfalls in technology and, chiefly, the lack of sufficiently powerful rocketry. With little prompting, McNutt can recite the long, dismal history of fizzled interstellar missions, listing study after study from across the decades (many of which he participated in) that all ultimately failed to launch.

The situation began to change with NASA's development of the Space Launch System (SLS), a wildly expensive and long-delayed megarocket that is slated for an inaugural test flight in 2021 and is planned to offer almost twice the thrust of any other booster now in operation. Although primarily meant to transport astronauts to the moon and Mars, the SLS could also hurl hefty robotic payloads at high speed throughout the solar system—assuming scientists make convincing cases for that pricey necessity (the latest estimates forecast SLS rockets may only fly once per year, at a per-launch cost of more than $2 billion).

NASA's premier science application for SLS has been the Europa Clipper mission, intended to seek out signs of habitability and life on an enigmatic, ocean-bearing moon of Jupiter. But the rocket's schedule slips may force Europa Clipper to launch on a smaller, slower, cheaper commercial booster instead. Such options are harder to contemplate for Interstellar Probe, whose need for speed would push even the mighty SLS to its limits (and thus help justify the rocket's existence).

"It's very dangerous to think that something better is coming later and to hope everything perfectly aligns," says Rob Stough, SLS's payload utilization manager at NASA's Marshall Space Flight Center. "It's great that the SLS is coming online and will be available. This is something that should be taken advantage of."

For now, McNutt and his colleagues concur. "This is right on the edge of what is technologically doable, and it doesn't look as good as what's in *Star Trek,* but I don't know where to go to buy those ships," he says. "[SLS] has a high price tag, but I know where to go to buy the darn thing."

To that end, NASA's heliophysics division—which, along with the earth science, planetary science and astrophysics divisions makes up the four pillars of the agency's Science Mission Directorate—is providing McNutt and his colleagues with $6.5 million across the next three years to firm up the science and engineering details for their notional mission. The resulting study would feed into the Heliophysics Decadal Survey in 2021, a once-every-10-years community assessment meant to guide the federal government in setting the nation's space-science goals and budgets. Topping the survey's recommendations when they are released in 2023 would be the biggest step yet toward making Interstellar Probe a reality. And failing to meet that mark would likely relegate it to at least another decade in mission-planning limbo.

Even if Interstellar Probe gains the survey's coveted blessing, however, McNutt and his colleagues believe it still needs "buy-in" from stakeholders elsewhere in NASA, such as the agency's planetary science and astrophysics divisions, to maximize its chances of actually flying. Hence the mission's packed provisional agenda of interdisciplinary science observations on its way out of the solar system, which run the gamut from studying dwarf planets to gathering the light of distant galaxies. Ideally, each set of observations would use its own dedicated instruments and rely on a unique, individualized flight profile—both impossibilities on this fast boat out of the solar system, where everything onboard must be multipurpose and even minor trajectory tweaks can carry major consequences.

The resulting tensions between different research communities hoping to capitalize on the mission were already apparent at a recent workshop dedicated to the Interstellar Probe's science held at the Explorers Club in New York City. To some attendees, the interdisciplinary discussions made the mission's supposed heliophysics focus seem to be almost an afterthought.

"If the main objective is to know the shape of the heliosphere, that should override anything else," says workshop participant Tom Krimigis, a veteran principal investigator of Voyager at APL who has worked on missions to Mercury, Pluto and every planet in between. "We must not make Interstellar Probe a 'Christmas tree.' If everyone gets to hang their own ornament, then, of course, it will be too heavy and costly and will sink of its own weight and never happen."

A key point of contention is where the spacecraft should actually go. Following in the Voyagers' footsteps by exiting in the vicinity of the heliosphere's thin, leading hemisphere—its "nose"—would presumably grant the quickest access to the interstellar medium and also allow closer studies of a mysterious ribbonlike feature of energized atoms first observed in 2009 that is draped across part of the heliopause. Moving laterally out of the solar system—out of the heliosphere's "flank"—could provide a better overall view of the heliosphere's shape and bolster studies of interstellar dust swept along at its turbulent edges. Another factor in favor of the flank is a potential plan for China's space agency to conduct a heliospheric mission of its own, proposed by Qiugang Zong of Peking University. Zong's proposal calls for twin "interstellar express probes," with one launched toward the nose and the other toward the tail as early as 2024—with both reaching around 100 AU by 2049.

"It would be interesting to see those results," says Kathleen Mandt, Interstellar Probe's deputy project scientist at APL. "And if we go out the flank, we'd have all three directions covered."

Scientists studying the swarms of dwarf planets and other icy flotsam beyond Pluto were perhaps the only participants at the workshop who were agnostic about the probe's trajectory. "Whichever way the heliophysics community wants to point the spacecraft, there's one or more very interesting [objects] in the way," says William McKinnon, a planetary scientist at Washington University in St. Louis. That agnosticism could end, however, if astronomers soon discover a fifth giant planet that circumstantial evidence suggests may lurk far out in the solar hinterlands. Leading theoretical models based on the putative planet's influence on the orbits of smaller objects suggest it may be five times the mass of Earth and somewhere between 400 and 500 AU away from the sun, toward the heliosphere's tail.

Such a world could be an irresistible target for Interstellar Probe as the spacecraft leaves the solar system.

"If a fifth giant planet were discovered, there would be a very strong case for going toward it," says Kirby Runyon, a researcher at APL who is leading planetary science planning for Interstellar Probe. "There would still be heliophysics to do with that sort of flyby. Maybe we should let the tail wag the dog."

Sending Interstellar Probe to fly by a newfound giant world would almost certainly cement the involvement of NASA's planetary science division, potentially boosting the mission's budget and securing its path to the launchpad. It would also, however, risk scuttling the mission's primary, heliophysics-driven science objectives. "If the heliosphere is shaped like a wind sock, do we get to the interstellar medium at all if we go toward the tail?" Provornikova says. "Some models predict the tail might extend for several thousand AU. If that's true, the spacecraft might get there eventually, but none of us would live to see it."

A Voyage In Space—And Time

Truth be told, even if it does launch, actuarial tables and the long timescales of deep-space missions offer bleak prospects for all but the very youngest of Interstellar Probe's planners to see the entire endeavor through. But this, too, is something being accounted for: last year, Mandt formally invited Janet Vertesi, a Princeton University sociologist who studies spacecraft teams, to take part in the Interstellar Probe mission.

"Most of the people planning this will be dead by the time it is finished," Vertesi says. "And none of the missions we've ever launched before have taken that kind of longevity into account. But Interstellar Probe needs to plan for a multigenerational, sequential process from the beginning. The baby boomers run it now and the Gen Xers will run it next. But millennials will be running it when it hits the heliopause, and Gen Z is going to run it when it reaches the interstellar medium."

Some of Vertesi's concerns are technical: how to structure scientific observations across multiple generations; how to read and preserve data from a half-century-old spacecraft. But the ones she ponders the most are cultural: How do you ensure the hard-won operational knowledge of one generation is passed on to the next and endures the assault of time? How do you guarantee individuals will gracefully step away after devoting their entire working lives to some greater, still unrealized cause? Solutions for those problems, she suspects, will not be found in any spreadsheet or supercomputer calculation. Successfully navigating beyond the solar system's limits will instead require rekindling the timeless arts of storytelling and of ritual, creating a mission-based mythos, around which successive generations of scientists can cohere.

"We are very deliberately preparing this mission not just to succeed in reaching the heliosphere but to also succeed in handing off leadership from one generation to the next," Paul says. "We realize that we are not the reasons for this mission—we are instruments through which the Interstellar Probe will be realized."

Building and Sending Interstellar Probes

7.0 Interstellar Probes Being Planned

7.1 Precursor Mission spacecraft

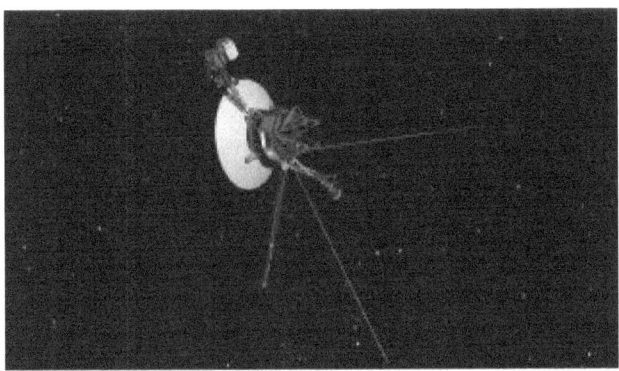

NASA's Voyager 1 spacecraft against a backdrop of stars. New ideas for a robotic interstellar mission are percolating. Ambitious science and strategic plans are being formulated for the fastest flight ever to interstellar space — almost six times faster than NASA's record-holding Voyager 1 spacecraft, which launched in 1977 and went interstellar in 2012.

With the goal of reaching 90 billion miles (145 billion kilometers) from the sun, the proposed robotic explorer would push the limits of engineering know-how and space technology, advocates say.

The Johns Hopkins Applied Physics Laboratory (APL) in Laurel, Maryland, is leading an international look at this prospect with a team of scientists and engineers studying a mission to the virtually unexplored space beyond our sun's sphere of influence.

"Overall, I think the study is progressing well and will provide some good and solid input for the next Decadal

Building and Sending Interstellar Probes

Survey round," said APL's Ralph McNutt, Interstellar Probe study leader and principal investigator. The Decadal Survey is based on studies led by the U.S. National Academies to provide a science community consensus on new undertakings in NASA space science and exploration.

NASA's twin Voyager probes have sailed beyond our solar bubble into interstellar space.

The subjects of interstellar travel, interstellar probes and interstellar "precursor" missions are not new, the paper explained, "but have lacked traction with policymakers and the scientific community at large because of the states of both scientific knowledge and engineering realities." Moreover, the paper's authors explained that the next step in reaching to the stars will require the recognition of engineering limits, scientific trades and scientific compromises, "but this is new neither in science nor exploration. Such a step would be an 'Interstellar Probe.' The time for that step has come."

The successful launch of NASA's groundbreaking Parker Solar Probe in August 2018 provides one more step forward for a realistic look at all of the possibilities, McNutt and his colleagues said. In addition, new discoveries in the Kuiper Belt by NASA's New Horizons mission — which flew by Pluto in 2015 and the even more-distant small object Ultima Thule last year — underscore the possibilities of compelling science as motivation for an Interstellar Probe, the researchers said.

Outward bound: NASA's New Horizons spacecraft flew by Pluto in July 2015, then cruised by the even more-distant object Ultima Thule on Jan. 1, 2019.

Building and Sending Interstellar Probes

Learn by doing

Current thinking about going interstellar begins with an "interstellar precursor" mission — a spacecraft capable of traveling to perhaps 1,000 astronomical units (AU) using current and near-term technology. (One AU is the Earth-sun distance — about 93 million miles, or 150 million km). This approach is a good one, both because of its potential science return and its impact on driving propulsion, communications and sensor technologies forward, said Paul Gilster, writer and editor at the website Centauri Dreams, which discusses ideas for the exploration of deep space, including the interstellar realm.

"We learn by doing, and pushing well beyond the heliopause will teach us what to expect in the local interstellar medium, where future, faster missions will eventually explore inner Oort Cloud objects like Sedna and push on into the Oort itself," Gilster said. (The Oort Cloud, which is home to trillions of comets, lies between about 1,000 and 100,000 AU from the sun.)

The huge reaches of space beyond the heliopause — the bubble blown by the solar wind — are terra incognita, he told Space.com.

"We can't even think of missions to more distant targets without mapping the immediate terrain to learn about hazards that could affect equipment, disrupt communications or even destroy the spacecraft," Gilster said.

An interstellar precursor is a logical successor to New Horizons and the Voyagers that have preceded it, he added.

Building and Sending Interstellar Probes

"It would be our first spacecraft truly designed for exploratory operations in interstellar space and would represent our intention to reach targets that today seem impossible, including one day the nearest stars," Gilster said.

The APL study — which focuses on a mission that could launch before 2030 and reach 1,000 AU in 50 years — is based on the next extension of what we know we can do, propulsion physicist Marc Millis, founder of the Tau Zero Foundation, said.

"It is a reasonable candidate for the next deep-space mission," Millis told Space.com. "It is not, however, a true interstellar mission. It is better referred to as an 'interstellar precursor' mission."

Voyager 1 has traveled about 142 AU during its four decades in space. While the interstellar-precursor probe would explore much more distant realms, it wouldn't get anywhere near another star system, Millis stressed. One thousand AU is about 0.016 light-years, and the closest star to the sun, Proxima Centauri, lies about 4.2 light-years away.

Still, "in addition to the value of its destination-based science, it could be used to assess the challenges of an exoplanet probe," Millis said of the putative mission.
The APL work is one of many interstellar approaches, he added.

"Others aim for more ambitious missions that need varying degrees of technological advancements, from the seemingly simple space sails to the seemingly impossible faster-than-light space warps," Millis said.

Building and Sending Interstellar Probes

Philosophical thoughts

Planning out futuristic interstellar missions also conjures up some deep, philosophical thoughts. "We need to reach for extreme targets, because exploration has always defined our species," Gilster said. "Our voyages have crafted our aspirations."

A full-on interstellar mission has long been thought impossible, or at least impractical, he added, but we know now that technologies are becoming available that can make it happen.

"A successful mission to another stellar system could show us possibly habitable planets up close and help us put our own human experience in the context of billions of worlds that could support life," Gilster said. "Philosophically, we can't help ourselves. We have to know what's out there, and whether we are alone in the cosmos."

Gilster's bottom line: "A first interstellar mission will drive new breakthroughs in propulsion, with inevitable ramifications for travel in our own solar system as well. Culturally and scientifically, such a mission is inevitable. The only question is when."

Building and Sending Interstellar Probes

Building and Sending Interstellar Probes

7.2 Interstellar Probe (1999)

Solar sails work by converting the energy in light into a momentum on the spacecraft, thus propelling the spacecraft. Felix Tisserand noted the effect of light pressure on comet tails in the 1800s.

The study by the NASA Jet Propulsion Laboratory, proposed using a solar sail to accelerate a spacecraft to reach the interstellar medium. It was planned to reach as far as 200 AU within 10 years at a speed of 14 AU/year (about 70 km/s, and function up to 400+ AU. A critical technology for the mission is a large 1 g/m2 solar sail.

This great journey requires advanced propulsion, and the 200-kg Interstellar Probe is designed to use a 200-m radius solar sail to achieve a velocity of 14 AU/year. After exiting the heliosphere within a decade of launch, it would be capable of continuing on to ~400 AU. Interstellar Probe would serve as the first step in a more ambitious program to explore the outer solar system and nearby galactic neighborhood.

Interstellar Probe, 1999

In the following years there were additional studies, including the Innovative Interstellar Explorer (published 2003), which focused on a design using RTGs powering an ion engine rather than a solar sail. Another project in this field for advanced spaceflight during this period was the Breakthrough Propulsion Physics Program which ran from 1996 through 2002.

Later examples of solar sail-propelled spacecraft include IKAROS, Nanosail-D2, and LightSail. Near-Earth Asteroid Scout is a planned light sail-propelled mission. For comparison, the LightSail spacecraft uses a sail 5 micron in thickness, whereas they predict a sail with 1 micron thickness would be needed for interstellar travel.

7.3 Breakthrough Starshot

The Breakthrough Starshot project: Launching ultra-fast light-driven nanocrafts to Alpha Centauri

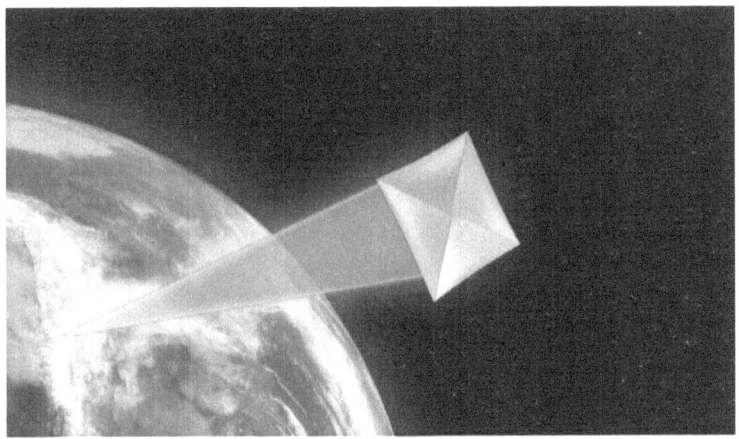

Harvard University's Professor Avi Loeb, Chair of the Breakthrough Starshot project, tells SciTech Europa about how the initiative plans to launch ultra-fast light-driven nanocrafts to Alpha Centauri within the next generation.

The Breakthrough Initiatives were founded in 2015 by Yuri and Julia Milner to explore the Universe, seek scientific evidence of life beyond Earth, and encourage public debate from a planetary perspective. The Breakthrough Starshot project aims to demonstrate proof of concept for ultra-fast light-driven nanocrafts, and lay the foundations for a first launch to Alpha Centauri within the next generation. Along the way, the project could generate important supplementary benefits to astronomy, including solar system exploration and the study of near-Earth asteroids.

Building and Sending Interstellar Probes

SciTech Europa Quarterly spoke with the Chair of the project's Advisory Committee, Professor Avi Loeb at Harvard University, USA, about the project's concept, design, technology, target, and future.

What is the rationale behind Breakthrough Starshot? Why was Proxima b chosen as the target?

The general goal of the project is to send a probe, including a camera, to the Alpha Centauri system, our nearest star system, in order to better understand what it is like there. When we first considered the project, we didn't know that there is a habitual planet (Proxima b) next to the nearest star, Proxima Centauri; we just imagined that this was a possibility and so the question was whether we could send a camera that would fly by that planet and take photographs so that we can tell if there is vegetation there, or whether there is an ocean; we wanted to see it from up close, rather than from a distance.

Of course, it is possible to gain some limited information from remote observations using telescopes, but that is very different from flying by and actually seeing the details. The main challenge that Proxima Centauri – or the Alpha Centauri system more generally – poses is the fact that it is four light years away, and if we want to get there within our life time, say in 20 years, then the spacecraft needs to move at a fifth of the speed of light.

The only method that seemed potentially feasible when we first started to consider this challenge was light sail technology, where, rather than carrying fuel, you use a powerful laser to push against a lightweight sail. We established the parameters for this about six months later, finding that we would need to use a very powerful laser (some 100 gigawatts in power) which will be delivered to the sail for a few minutes. If the sail weighs roughly a gram, and the payload (including the camera, navigation device, and communication device) weighs roughly a gram, then it is possible to reach a fifth of the speed of light by pushing on the sail, which is a few metres in size, with the laser.

The sail needs to be in place outside of the Earth's atmosphere, otherwise the friction would prevent it from working. As such, the concept sees the craft being released from a 'mother ship' above the atmosphere, with the laser being fired from the ground through the atmosphere to the sail. We have investigated the parameters of the system and have not come across anything which should prevent us from realizing this ambition. The amount of power required from the laser is roughly the same as that which is required deliver to the space shuttle into orbit. But, of course, because our craft weighs so little it is able to reach exceptional speeds. Even at a fifth of the speed of light, it would take about 20 years to reach the Alpha Centauri system, and then another four years for the signal to be communicated back to us on Earth. Nevertheless, this will be the very first time that humans have been able to bridge the gap between stars, which will be a major milestone.

The first phase of the project will take place over the next decade or so, and we will be focusing on the technology development and investment in the most critical technologies that are essential for this concept to work, such as the laser beams. We will also be working to develop the materials for the sail. In terms of material selection, we have already done some preliminary work and have also made some progress towards optimizing the shape of the sail, although there is still work to be done. The sail needs to ride on the laser beam in a stable fashion. These are the two critical technologies we are starting with. After that, we will progress into the communication technology and other issues.

We have already selected about a dozen proposals from experimental teams that are able to work on the laser technology, and a similar number will work on the sail technology. Funding for this work has now begun to be allocated.

How will the sail differ from existing solar sails?

A solar sail follows a similar concept, in that it is lightweight and is pushed by light. Although, of course, in the case of a solar sail this light comes from the Sun. Sunlight, however, has a limited amount of power, perhaps a kilowatt per meter, while with our laser it is possible to reach a much stronger power per unit area, not least because the material has been tuned to fit the frequency of the laser. A second advantage of using a laser is also the fact that the amount of power you can put on a given area is under your control. You can also tailor the shape and time evolution of the laser beam, such that it will optimize the launch. The power for the laser will ultimately come from the Sun because we envisage it being powered by energy generated from solar cells.

Given that this is a long-term initiative, how do you hope to be able to build on the future discoveries set to be made, for example, by the JWST?

Of course, any scientific information about the Alpha Centauri system, especially Alpha Centauri a and b and Proxima Centauri, will be very helpful. Proxima Centauri is a dwarf star that is just 12% of the mass of our Sun, and Proxima b is a planet located 20 times closer to its star than the Earth is to the Sun. Despite this relative proximity, the planet is potentially habitable because the Proxima Centauri is much fainter than the Sun, meaning that it needs to be 20 times closer in order to reach a similar surface temperature to Earth.

However, we only know that this planet exists; we don't know if it has an atmosphere, which is necessary to support life because otherwise water ice will instantly become a gas and evaporate into space – which is why Mars doesn't have liquid water on its surface. As such, it is important for us to find out whether Proxima b has an atmosphere. One way to do that will be with the James Webb Space Telescope.

It is possible to tell, without visiting the planet, that Proxima b is tidally locked to its star and thus that there is a permanent day side and a permanent night side, with a temperature contrast between them because the same side of the planet always faces the star. The permanent day side is much hotter than the night side, and that temperature contrast can inform us about whether there is an atmosphere because an atmosphere would moderate the temperature difference. For instance, there would be rain, and if there was an ocean, that would moderate the temperature even further.

We would also like to know just how much hotter the day side is relative to the night side, and that can be determined by observations with the James Webb Space Telescope because it will be able to look at the light of that planet as it moves around the star and, by analyzing the light curve, we should be able to discern the level of temperature contrast.

By modelling that light curve, moreover, it will be possible to compare that to an environment, for example, of bare rock, and so gain some potential insights as to what the surface of the planet is like. The James Webb Space Telescope thus has the potential to provide us with evidence supporting the idea that there is an atmosphere on the planet and, of course, if this proves to be the case then it will become a much more attractive place for us to visit with Starshot.

There is also the potential of finding other planets in the habitable zone around Alpha Centauri a and b and, with future telescopes such as JWST there will be teams working to identify such planets, both in this system and elsewhere, which is another exciting thing to look forward to in the next decade.

Our plan for Starshot also involves building a dedicated space station to look for habitable planets around Alpha Centauri a and b. As such, moving forward we stand to learn a lot more about Proxima b in terms of its habitability, whether it has an atmosphere, and whether there are more planets in the vicinity which we would like to visit.

Does that mean that the target is relatively flexible, in that should the JWST etc. provide information to suggest that there is no life on Proxima b, but that another planet elsewhere may have, then you can switch targets?

Absolutely. In principle, the spacecraft are relatively cheap, with most of the investment being spent on the infrastructure (the laser and launch systems). Once that is in place, that spacecraft would only cost in the region of $100 (~€88) each, and so the idea would be to launch a number of them (perhaps one per day or one every few days) and we could potentially aim them at different targets, with some visiting other planets and systems as they travel to their final destination. And, given the speed at which these spacecraft – which will be the size of a typical mobile telephone – will be travelling, they would be able to reach a number of interesting targets within the Solar System in relatively short timeframes.

For instance, it would take just a few days to reach Pluto, rather than the nine and half years that it took NASA's New Horizons mission. We could also investigate the existence of so-called 'planet nine', which is believed to exist at the edge of the Solar System, or we could fly by other objects of interest and take photographs of them. For example, the first interstellar asteroid was recently discovered travelling through the Solar System. However, even if chemical rockets had been used it would have been impossible to chase 'Oumuamua' down and take photographs of it. However, had Starshot been operational, we could have reached it quite easily. The Breakthrough Starshot laser beam could also be used to ablate and hence deflect a dangerous asteroid.

In your last article you argued that there is a need for more investment in building better observatories and searching for a wide variety of artificial signals in the sky. How would you like to see this being approached?

The search for extra-terrestrial life is becoming a very prominent theme in space-based plans for the future, particularly in the context astrobiology, which is the combination of biology and astrophysics, but also in the mainstream, with projects and missions searching for signatures of primitive and microbial life, for example, in the composition of a planet's atmosphere. If we can infer the composition to contain oxygen, for instance, or methane, then that could be indicative of microbial life.

At the same time, one can imagine searching for intelligent, technological civilizations by looking for 'techno-signatures'. We already have the technology to search for such signals in space, including the industrial pollution of atmospheres, artificial lights, the redistribution of heat, mega-structures, and so on. It could also be possible to discover the burned surfaces of planets which have seen devastating changes in their climate as a result of the activities of a past or current civilization, or perhaps the remains of other civilizations which no longer exist. This could form the basis of 'space archaeology', where people search for the signatures of civilizations that are now extinct.

The search for both primitive and technologically advanced extra-terrestrial life should be pursued because we know that humans exist on Earth, and we know that about a quarter of all the stars in the Milky Way galaxy have a planet the size of the Earth orbiting them which has a surface temperature similar to Earth's, meaning that liquid water may exist of their surface, just like Proxima b, and so the chemistry of life as we know it may be there. The realization that Earth-like planets are so common should push us to search for either microbial or technological life, and I think the next decade will be very exciting as people start to do so. If we are successful, then this would answer one of humanity's most fundamental questions: 'are we alone?'

What are your short-term goals for Starshot?

In the first instance, we want to demonstrate the technologies, and this will constitute the research and development phase over the next decade as we work to demonstrate that we are able to combine numerous small lasers into a coherent laser beam that is powerful enough for our task. We will also be working to demonstrate that we can build a lightweight sail made of a material that is strong enough to sustain the forces acted upon it by the laser and, indeed, that is designed in a shape that can ride, in a stable fashion, on the laser beam. Then, we want to be able to demonstrate that we can communicate back to Earth across the vast distances that the project will involve.

These are significant challenges, but they are not insurmountable, and people are very excited about what we hope to achieve, especially the general public, because since the Apollo missions there have been very few others to demonstrate such a level of ambition. Going to another star for the first time is much more challenging than many of the other missions we have seen in recent years, and it is exciting because it could potentially help us understand our place in the Universe. If we find a technologically advanced civilization, of course, then that would change everything; we could learn from them, we could see advanced technologies that we haven't even dreamed of. Developing these technologies will also have a lot of benefits for other applications.

Where do you think those are going to lie moving forwards?

The impact on astronomy will be tremendous because, at the moment, the field is focused on the physical Universe in the sense that we imagine a Universe populated with stars and planets that are totally devoid of life. If we do find any evidence of life, either microbial or technological, then this will entirely change our perception of the sky.

When I look up at the stars in the sky at night, it is easy to think of them as just the lights from distant planets and stars. But if I imagine that there could be another being out there looking back at me, then that changes everything; not least my own perception of my place in the Universe.

Finding life on another planet would have a profound impact on human psychology, too, as well on philosophy and religion, and not just on the science of astronomy. In a general sense, it is not too much to presume that this discovery would also have an impact on the way people behave and how they interact with each other here on Earth, because we might come to feel as though we are a part of a single, unified team, humanity, and stop focusing so much on mundane issues like geographical borders.

The other thing to keep in mind is that, one way or another, humans will have to leave Earth at some point. The latest we can imagine staying here is for another billion years or so, when the Sun will heat the Earth up to the point that the oceans will boil. But before that, there could be an asteroid impact of the type that killed the dinosaurs, while climate change or a nuclear war could leave the Earth uninhabitable.

Currently, should any of those things happen, then we risk extinction as we are unable to leave the Earth for other planets in any real meaningful way. As such, we need to start exploring how we are going to stretch out into space and, in order to do that, we must start at the beginning with small spacecraft so that we can demonstrate that it is indeed possible to reach very great distances. That is just the first step; beyond that, we will start exploring how to move larger objects at faster speeds.

You cannot start the journey without the first steps. Even the critics who argue that the necessary technologies don't exist yet will admit that you have to start somewhere; you have to have a dream. Otherwise, we will never leave the Earth and we will be left to our fate.

Building and Sending Interstellar Probes

Building and Sending Interstellar Probes

8.0 Thinking About Probe Designs

Lessons on Designing an Interstellar Probe

Engineering full-scale interstellar probes will likely require new materials, new tech and a whole new set of design parameters. But most of all it will require speed.

Today's fastest spacecraft --- NASA's New Horizons, Voyager 1 and Voyager 2 --- all still only travel at a fraction of one percent of light. And even at one percent of light speed, it would take 400 years to get to the next star over. Therein, lies the rub.

"There's a well-known problem in deep space travel, the Wait equation," Jeff Greason, Chairman of the Board for The Tau Zero Foundation, a non-profit group currently working on interstellar and advanced propulsion technologies, told me. "The value of investing in something today that won't return science for 400 years is very low." It makes no sense to launch a probe to another star if the spacecraft is traveling at only one percent of light , says Greason. That's because by the time the 400-year probe got there, he says, it would have long been passed by faster, later probes. That's one reason The Tau Zero Foundation is focusing on improving propulsion technologies first.

"It's far easier to make an interstellar mission take twenty years than it is to figure out how to build, and how to fund a 200-year mission," said Greason.

But there are also other challenges.

"The real challenges are in the lack of maintenance and redundancy, in protecting the craft from erosion by interstellar dust and gas at such high speeds," said Greason.

There's also the need to work out how to make such craft both autonomous and capable of communications back to Earth.

An antenna sending data back to Earth from interstellar distances will require autonomous pointing stability, Ralph McNutt, chief scientist for space science at Johns Hopkins University's Applied Physics Lab in Maryland. He says it would also require a really good clock to predict where Earth will be once the interstellar craft's downlink signal arrives. At 1000 AU (Earth-Sun distances), that's already about 5 light days out, says McNutt. Thus, McNutt says the spacecraft's transmitting antenna must be pointed to where Earth will be five days after it downlinks its communications.

Building and Sending Interstellar Probes

"Lasers can be more tightly focused than the microwave frequencies used by missions today," said Greason. "It's also possible to use a receiver out at the thousand AU kind of distance as a relay."

The first interstellar missions will not be to the Alpha Centauri system, the next stars over, roughly some four light years away. Rather, they will be interstellar precursor missions ranging from 200 to 40,000 AU.

Our Sun is in the Local Interstellar Cloud, but Alpha Centauri is in a different composition of gas and plasma called the G Cloud," Greason says.

"The nearest edge of the G cloud is about 40,000 AU away," said Greason. "So a mission to sample it would be a very useful precursor to an Alpha Centauri mission." In fact, the first probes to truly go interstellar may be gram-scale robotic nano-craft of the sort advocated by the Breakthrough Starshot Initiative. They advocate launching a lightsail mothership into high Earth orbit which would then be powered by beamed light energy from the ground. Each lightsail would contain thousands of so-called 'StarChips' nano-craft.

Building and Sending Interstellar Probes

Once these StarChips reach their interstellar destination, says Greason, their sheer numbers upon deployment would ensure redundancy and mission success. The initiative notes that each starchip would be equipped with miniaturized cameras, photon thrusters; even navigation and communication equipment.

For larger craft, there's the idea of a plasma magnet as advocated by John Slough, a professor of aeronautics and astronautics at the University of Washington in Seattle. The idea boils down to harnessing the solar wind in such a manner as to generate unprecedented speeds.

Interplanetary and interstellar space is full of a thin plasma of charged particles, says Greason. A rotating magnetic field pulls those particles around, making a loop that he says is the source of a 'steady' magnetic field. Even as the solar wind becomes weaker, Greason says, this plasma magnet can expand to provide the spacecraft it's propelling with constant thrust.

Greason says in theory such plasma magnets can be used as magnetic sails to reach speeds of 400 kilometers per second without the need for propellant. Proof of technology missions to demonstrate this concept, says Greason, could in principle be done for tens to hundreds of millions of dollars.

But such tests remain years away as does much of the architecture to get us to Alpha Centauri.

Even so, as McNutt points out: "The Wright brothers did not throw up their hands and go back to bicycles full time because they had no concept of an SR-71 Blackbird."

Building and Sending Interstellar Probes

Building and Sending Interstellar Probes

9.0 Summary

It seems sure that the launch of an interstellar probe to visit the nearest star will happen in the twenty-first century. The plans will also be to accelerate it fast enough that we can see the results of it reaching another star system (Most likely Alpha Centauri) within fifty years. This would mean many people alive when it launched would also be here when we hear back from the probe.

I was surprised from my research on the subject that engineers and scientists have been thinking about interstellar probes seriously for fifty or sixty years.

There are also some approaches to designing and building a probe to visit Alpha Centauri in our lifetimes by sending very fast probes.

These ideas include a solar sail and very small probe boosted by an array of lasers to allow it to reach relativistic speeds.

I hope one of these probes is launched in my lifetime so we can start seeing real pictures and data from another star and maybe also pictures and data from an Earth like planet around that star.

All the Best,

Martin K. Ettington

July 2020

Building and Sending Interstellar Probes

10.0 Bibliography

1. Interstellar Probe.
*https://en.wikipedia.org/wiki/Interstellar_probe#:~:text=Ther
e%20are%20five%20interstellar%20probes,three%20are%
20on%20interstellar%20trajectories.* [Online]

2. Nasa Interstellar Probe Mission.
*https://www.space.com/42935-nasa-interstellar-probe-
mission-idea.html.* [Online]

3. Interstellar Concepts.
*https://en.wikipedia.org/wiki/Interstellar_probe#Interstellar_
concepts.* [Online]

4. Interstellar Mission Reaches for the Stars One
Generation at a Time.
*https://www.scientificamerican.com/article/proposed-
interstellar-mission-reaches-for-the-stars-one-generation-
at-a-time1/.* [Online]

5. Proposed Interstellar Mission Reaches for the Stars on
Generation at a time.
*https://www.scientificamerican.com/article/proposed-
interstellar-mission-reaches-for-the-stars-one-generation-
at-a-time1/.* [Online]

6. Interstellar Probe (1999).
https://en.wikipedia.org/wiki/Interstellar_Probe_(1999).
[Online]

7. Breakthrough Starshot.
https://en.wikipedia.org/wiki/Breakthrough_Starshot.
[Online]

8. Ion Thruster. *https://en.wikipedia.org/wiki/Ion_thruster.*
[Online]

9. Solar Sail. *https://en.wikipedia.org/wiki/Solar_sail.* [Online]

10. Oberth Effect. *https://en.wikipedia.org/wiki/Oberth_effect.* [Online]

11. The Breakthough Starshot Initiative. *https://www.scitecheuropa.eu/alpha-centauri/93771/.* [Online]

12. Getting Real About Interstellar Probes. *vhttps://www.forbes.com/sites/brucedorminey/2017/06/19/ getting-real-about-interstellar-probes/#57984c6b1576.* [Online]

Building and Sending Interstellar Probes

11.0 Index

www.ingramcontent.com/pod-product-compliance
Lightning Source LLC
Chambersburg PA
CBHW030954240526
45463CB00016B/2549